THE ANIMAL KINGDOM

Arthropods

Bev Harvey

CHELSEA CLUBHOUSE

An Imprint of Chelsea House Publishers
A Haights Cross Communications Company

Philadelphia

This edition first published in 2003 in the United States of America by Chelsea Clubhouse, a division of Chelsea House Publishers and a subsidiary of Haights Cross Communications.

Chelsea Clubhouse
1974 Sproul Road, Suite 400
Broomall, PA 19008-0914

The Chelsea House world wide web address is www.chelseahouse.com

Library of Congress Cataloging-in-Publication Data

Harvey, Bev.
   Arthropods / by Bev Harvey.
   p. cm. — (The animal kingdom)

   Summary: Introduces the physical characteristics and habits of various types of arthropods including insects and spiders as well as lobsters, crabs, and shrimp.

   ISBN 0-7910-6986-9
   1. Arthropoda—Juvenile literature. [1. Arthropods.] I. Title.
   QL434.15 .H37 2003
   595—dc21

                                                                              2002000921

First published in 2002 by
MACMILLAN EDUCATION AUSTRALIA PTY LTD
627 Chapel Street, South Yarra, Australia, 3141

Copyright © Bev Harvey 2002
Copyright in photographs © individual photographers as credited

Edited by Angelique Campbell-Muir
Page layout by Domenic Lauricella

Printed in China

Acknowledgements

Cover photograph: Coral crab, courtesy of ANT Photo Library.

ANT Photo Library, pp. 1, 4, 5, 9, 11, 12, 16, 20, 22, 25; Kathie Atkinson/Auscape, pp. 14 (eggs & adult butterfly), 24; Jean-Paul Ferrero/Auscape, p. 6 (center); François Gilson/Bios Auscape, p. 28; P. Goetgheluck—PHO. N. E. /Auscape, pp. 7 (center), 15; Karen Gowlett-Holmes—Oxford Scientific Films/Auscape, p. 17; Greg Harold/Auscape, p. 26; Wayne Lawler/Auscape, p. 14 (caterpillar); Reg Morrison/Auscape, pp. 6 (top), 13; D. Parer & E. Parer-Cook/Auscape, p. 7 (bottom); Fritz Polking/Auscape, p. 14 (adult butterflies mating); John Shaw/Auscape, p. 14 (cocoon); Australian Picture Library/Corbis, p. 19; Jason Edwards/Bio Images, pp. 7 (top), 27; Dennis Crawford/Graphic Science, p. 18; Peter Marsa Arthur Wigley/Royal Melbourne Hospital Medical Illustrations Department, p. 23.

Contents

Arthropods

Arthropods are animals such as ants, spiders, and crabs. Arthropods are the largest group in the animal kingdom. Arthropods are invertebrates. An invertebrate does not have a skeleton inside. Instead, an invertebrate has a shell called an exoskeleton on the outside of its body. Arthropods are **cold-blooded**.

An arthropod's body is divided into segments, or sections. Body parts such as wings, legs, and **antennae** are attached to the segments. These segments and body parts are jointed so an arthropod can move.

Types of Arthropods

There are many types of arthropods.

Scorpions live in warm, dry areas such as deserts. They have poisonous stingers on their tails.

Centipedes live under stones or wood. They have many body segments and many legs.

Spiders live all over the world. They have two body segments and eight legs.

Ants are also found around the world. They live in large **colonies**.

Butterflies live on every continent except Antarctica. Butterflies are active during the day.

Most crabs live at the bottom of oceans. Some come onto land, and others live in fresh water.

Insects

Insects are arthropods. Beetles, moths, and flies are insects. Scientists have found more than 1 million kinds of insects. An insect usually has three body segments. It has a head, thorax, and abdomen. An insect usually has antennae attached to its head. It has six legs attached to its thorax.

Many insects can fly. They move their wings very fast stay in the air. An insect's wings are attached to its thorax. Most insects are less than one-fourth inch (6.4 millimeters) long.

Arachnids

Spiders, scorpions, ticks, and mites are types of arachnids. Arachnids are another kind of arthropod. An arachnid usually has two body segments. The cephalothorax contains its head. Eight legs attach to its cephalothorax. An arachnid uses its front legs to catch its **prey**. The abdomen is its tail section.

Crustaceans

Lobsters, crabs, and shrimps are kinds of crustaceans. Crustaceans are arthropods, too. A crustacean usually has three body segments. It has a head, thorax, and abdomen. A crustacean usually has antennae, eyes, and a mouth on its head. Its legs attach to its thorax. Some crustaceans have pincers, or claws, for catching food. The abdomen is the crustacean's tail.

Eating Habits

Arthropods eat a variety of foods. Many insects are herbivores. Herbivores eat plants. Insects such as beetles and grasshoppers have mouthparts for chewing. They can eat an entire plant. Insects such as aphids and cicadas have sucking mouthparts. They suck the juice out of plants.

These aphids are sucking the juice from a rosebush.

This spider has caught an ant in its web.

Some arachnids, such as spiders, are carnivores. Carnivores eat other animals. Spiders spin strong fibers called silk to make webs. Insects become caught in the sticky fibers of the web and cannot escape. The spider then catches and eats its prey.

Young Arthropods

Many arthropods lay eggs to create young.
Insects such as butterflies lay eggs. Butterfly
eggs hatch into caterpillars. A caterpillar eats
and grows, then it turns into a **pupa**. It
changes inside its hard shell. It breaks out as a
butterfly. This process is called **metamorphosis**.

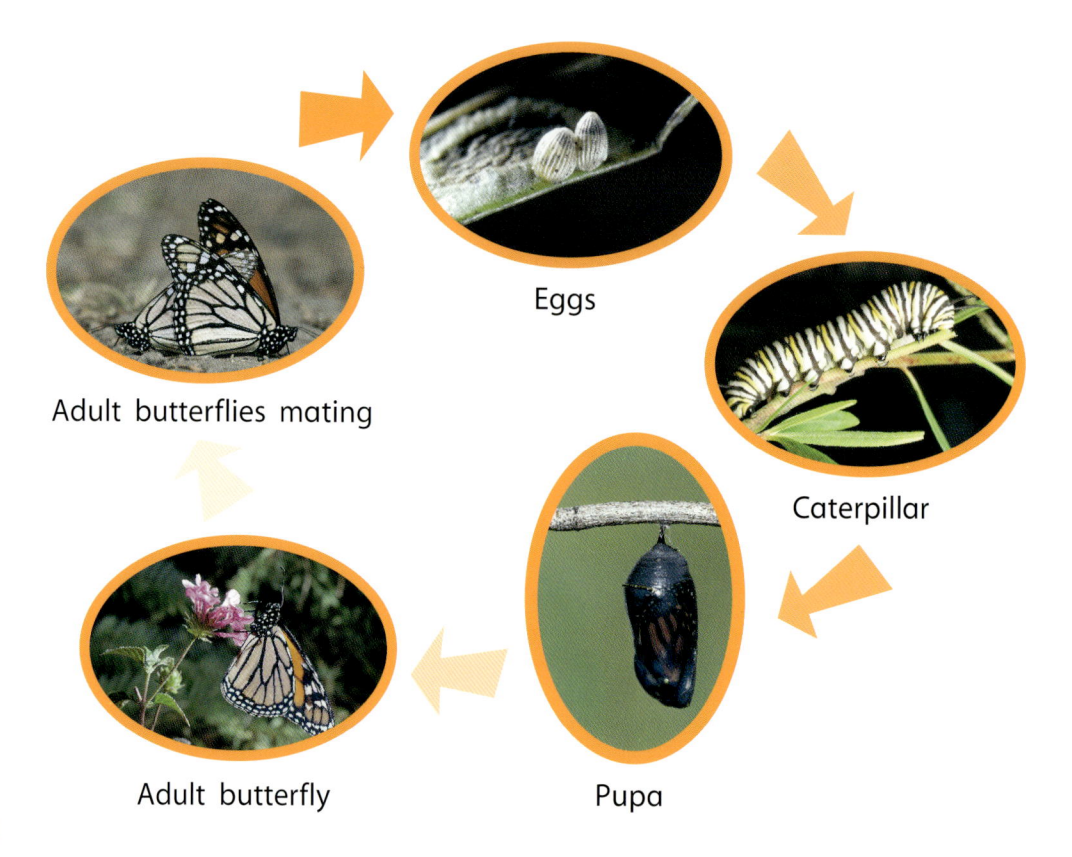

Eggs

Adult butterflies mating

Caterpillar

Adult butterfly

Pupa

Many arachnids also lay eggs. Some spider eggs are protected inside cocoons. **Larvae** hatch from the eggs, but they stay inside the cocoon. The larvae **molt** and change into young spiders.

Most female crustaceans lay eggs too. They carry their eggs with them until they hatch.

Pet Hermit Crabs

A hermit crab is a crustacean. Other types of crabs grow thick shells to cover their abdomens. The hermit crab does not. Instead, it lives in empty shells left by other animals.

Some people keep hermit crabs as pets. They need to provide shells for their hermit crabs. As a hermit crab grows, it needs to move into a larger shell.

A hermit crab without its shell.

A hermit crab grows by molting. When it molts, it loses its exoskeleton. It takes a few weeks for its new skeleton to harden. Molting usually happens twice a year.

17

Hermit crabs need a warm, humid environment. In the wild, hermit crabs live on land, so a pet hermit crab needs sand in its aquarium. It also needs fresh water to drink and salt water to bathe in.

Every day, a pet hermit crab needs to eat fresh foods, such as coconut and apple. It can also eat hermit-crab pellets from the pet shop.

Funnel-web Spiders

Funnel-web spiders live in parts of Australia. The funnel-web spider is poisonous to humans. Its bite can be deadly.

Funnel-web spiders can grow to about 2 inches (about 5 centimeters) long. A funnel-web spider lives alone. It burrows under a rock or inside a log. It makes a funnel-shaped web.

During summer and autumn, a male funnel-web spider leaves its burrow. It searches for a female to mate with it. This is when people are most likely to be bitten by a funnel-web spider. People become very sick and can die from funnel-web spider bites.

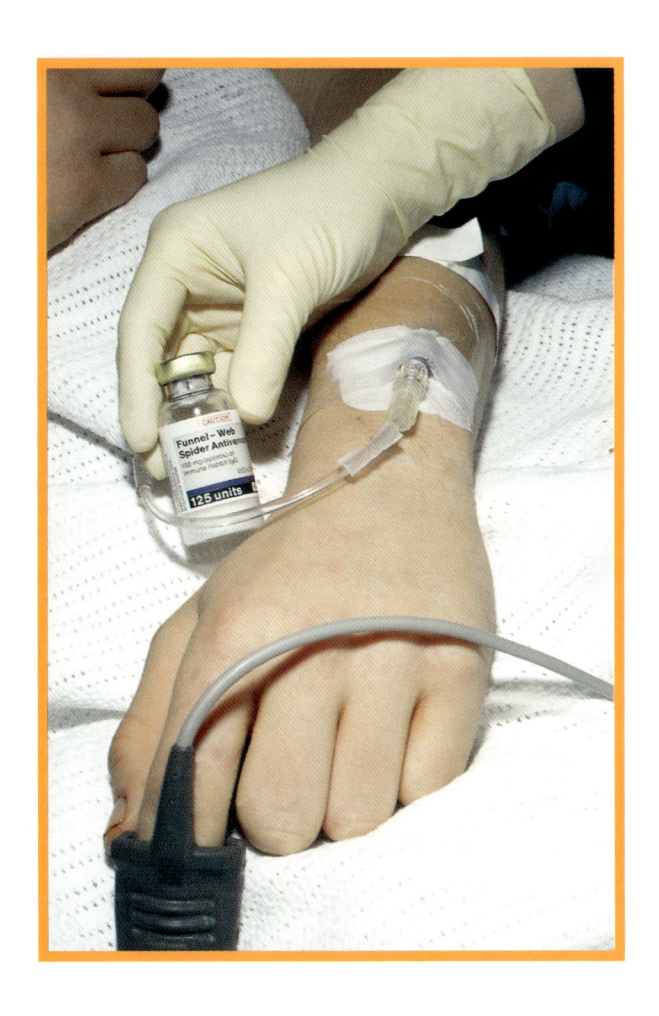

Fortunately, doctors have an antivenin for people bitten by funnel-web spiders. This medicine stops the spider's poison from working.

Ants

Ants are insects that live all over the world except extremely cold places. Most ants make nests in soil, sand, leaves, or wood. They live together in large groups called colonies.

Ants have different jobs. Many colonies have worker ants, male ants, and one queen ant. The workers gather food, care for the young, and build nests. Male ants mate with the queen. The queen ant lays eggs all her life.

Ants have two sets of jaws. They use their outer set for carrying food and their inner set for chewing food. Most ants are **omnivores**. They eat plants and other insects. Worker ants gather food for the colony.

The queen ant lays many eggs. The eggs hatch into white larvae. The larvae eat and grow. Then they turn into pupae.

The pupae are covered with silk cocoons or thin skins. The pupae change into adults. Then they come out of the cocoons or skins.

Endangered Queen Alexandra's Birdwing Butterflies

The **endangered** Queen Alexandra's birdwing butterfly is the largest butterfly in the world. It has a wingspan of up to 12 inches (30 centimeters). These butterflies are rare and live only in the rain forests of New Guinea.

These butterflies are endangered because the
rain forests where they live are being cleared
for logging and farming. People also catch
them and sell them to collectors. It is now
illegal to catch and sell Queen Alexandra's
birdwing butterflies.

Animal Classification

The animal kingdom is divided into two main groups of animals: invertebrates and vertebrates. In this book, you have read about arthropods.

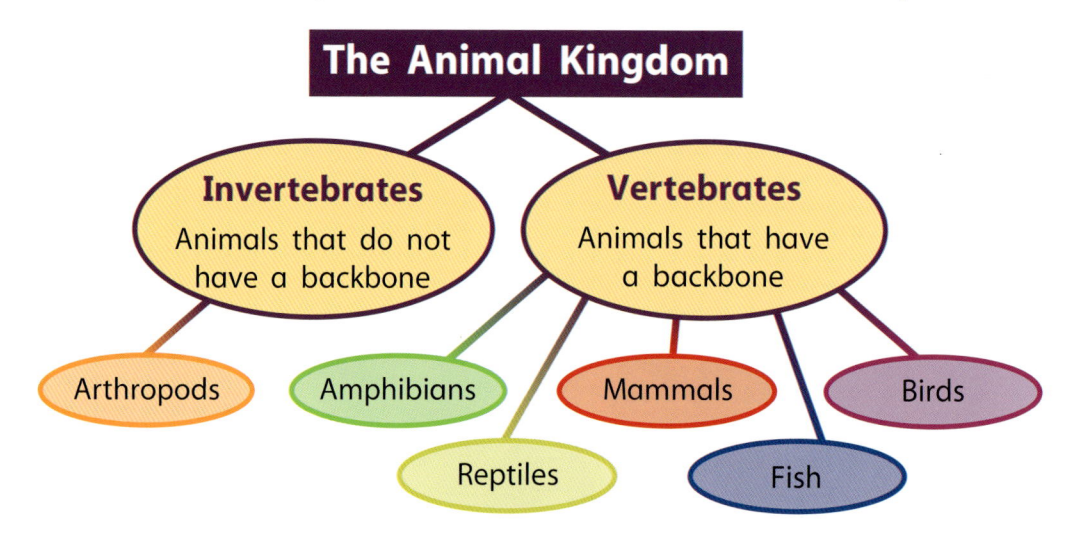

Question What have you learned about arthropods?

Answer Arthropods:
- are invertebrates
- have bodies with segments
- are the largest group in the animal kingdom.

Glossary

antennae a pair of long, thin body parts attached to the animal's head; most antennae are used for touching, but some are used for smelling.

cold-blooded an animal whose body temperature changes to match the temperature of the air, ground, or water around it

colony a group of the same kind of animal living together

endangered a type of animal or plant that may soon die out

larva a growth stage some animals go through after hatching from their eggs and before becoming adults; larvae look very different from adults; tadpoles are the larval stage of frogs; caterpillars are the larval stage of butterflies.

metamorphosis the change in the form of an animal as it grows to adulthood

molt to shed a skin or covering to uncover the new skin or covering underneath

omnivore an animal that eats both plants and meat

prey an animal hunted for food

pupa the inactive form of some animals between the larval and adult stages; pupae usually have hard coverings.

Index